Why Do Some Animals Hibernate?

Patricia J. Murphy

The Rosen Publishing Group's
PowerKids Press™
New York

To my family

Published in 2004 by The Rosen Publishing Group, Inc.
29 East 21st Street, New York, NY 10010

First Edition

Editor: Frances E. Ruffin
Book Design: Danielle Primiceri
Layout: Michael de Guzman

Photo Credits: Cover © Digital Stock; p. 4 © Ted Levin/Animals Animals; p. 7 © John Cancalosi/Peter Arnold, Inc; p. 8 © Marty Stouffer/Animals Animals; p. 11 © S. J. Krasemann/Peter Arnold, Inc.; p. 12 © Lynn Rogers/Peter Arnold, Inc; p. 15 © Leonard Lee Rue III/Animals Animals; p. 16 top © Corbis, middle © Richard Farnell/Animals Animals, bottom © Phillip Colla/Innerspace Visions; p. 19 © Richard T. Nowitz/Corbis; p. 20 © Joe McDonald/Animals Animals.

Murphy, Patricia J., 1963–
Why do some animals hibernate? / Patricia J. Murphy.
p. cm. — (The library of why?)
Includes bibliographical references (p.).
Summary: Explains what hibernation is, why and how some animals hibernate, what animals do in the winter if they don't hibernate, and how we can help animals survive the winter.
ISBN 0-8239-6232-6 (lib. bdg.)
1. Hibernation—Juvenile literature. [1. Hibernation.] I. Title. II. Series.
QL755 .M87 2003
591.56'5—dc21

2001005465

Manufactured in the United States of America

Contents

What Is Hibernation?

Animals hibernate when it is cold and snowy, and when there is little food. "Hibernate" comes from the **Latin** word *hiberna*. It means "winter." Hibernation is nature's way of taking care of animals in the winter. It is a deep sleep that helps animals to live through long, cold winters. Without it many animals would die. Raccoons may hibernate for a few weeks at a time. Woodchucks or groundhogs may hibernate for as long as seven months. Hibernation is an **adaptation**. It is a way that animals change to fit their **environments**.

◀ *Raccoons and other animals hibernate because food may be difficult to find during winter.*

How Do Animals Prepare to Hibernate?

As winter approaches, the **hypothalamus**, a part of an animal's brain, makes the animal aware of the shorter hours of daylight and the colder **temperatures**. This organ also tells animals to store food for the winter. Bears, brown bats, and woodchucks store food by eating extra food in the fall. This adds extra fat to their bodies. The stored fat helps animals to live through the winter months. Other animals, such as chipmunks and pikas, store food in their dens or in underground tunnels.

Pikas cut down green plants and then store them in their underground tunnels, which are called burrows. ▶

What Happens When Animals Hibernate?

While an animal hibernates, its body changes. Its body temperature becomes cooler. Its heart rate, breathing, and **metabolism** all slow down. This slowing down allows the animal to save energy in cold temperatures. Its body lives off its stored fat. The food that it eats during the fall becomes a type of fat called brown fat. This brown fat keeps the animal's body alive during winter. The brown fat also helps the animal to wake up. It **fuels** the animal's brain, heart, and lungs, and gives the animal energy to move.

As other animals have, this grizzly bear has an inner clock that tells it when to wake up and stop hibernating.

Which Animals Hibernate?

Some animals sleep all winter. They appear almost dead. Their metabolisms become very slow. They may not react to sound or touch. These "true hibernators" include small, **warm-blooded mammals**. Some larger mammals, such as bears and raccoons, sleep through winter. Their breathing, metabolisms, and heart rates remain normal. They are easily awakened. **Cold-blooded** reptiles and amphibians are also true hibernators. Snakes hibernate in holes in the ground. Turtles and frogs dig into the muddy bottoms of ponds.

This brown bat is covered with dew. When its body temperature drops too low, it wakes up briefly to bring up its temperature. ▶

Do Animals Wake Up During Winter?

All hibernating animals will wake up every few weeks. They get up to eat or to get rid of waste. Warmer weather also may stir them from their sleep. They may wake because they want to fill their stomachs! Their stored fat may not be enough to last through winter. During winter, baby bear cubs are born while their hibernating mothers are half asleep. The cubs live on their mothers' milk. When spring arrives, mother bears and their cubs leave the den to eat.

These young black bear cubs stay in a den with their hibernating mother to avoid becoming a hungry animal's winter snack.

How Do Animals Stay Active During Winter?

Many animals stay active through the entire winter. They have special **traits** that allow them to be active. These animals grow thick fur coats to keep them warm. Many find dead food, such as dead leaves, bark, or twigs, below the snow. Some may also hunt small animals. Deer, rabbits, some squirrels, and foxes are active winter animals. During winter storms some animals keep warm in the stumps of trees or under logs. Snowshoe rabbits grow long hair behind their toes. This turns their paws into wide "snowshoes."

White-tailed deer have long legs to make it easy for them to walk through the snow. ▶

What Is Migration?

Some animals migrate instead of hibernating. By migrating, or moving, to warmer places, they escape cold temperatures and the lack of food. Most return in the spring. Monarch butterflies migrate south to California, Florida, or Mexico, for warm temperatures. Many birds migrate south each year. The gray whale swims thousands of miles (km) from the freezing waters of Alaska to the warm waters of Baja, California, each year. While there female whales give birth to baby whales. The whales swim home to Alaska in the spring.

◀ *Snow geese in flight* (top left), *herds of caribou* (center), *and gray whales* (bottom) *all migrate to warmer places in winter.*

Do Humans Hibernate?

Some people who live where the weather turns cold may stay indoors during most of the winter. People do not sleep all winter. However, during heavy winter storms, they make sure to have enough food in the house. Scientists believe that early humans learned from animals about how to adapt to winter. People wore animal fur and wool to keep warm. They stored food for the winter. Today we still use what early humans learned from animals to get through the long, cold winters.

These people are gathering wood to keep warm during winter's cold weather. ▶

What Are Scientists Learning About Hibernation?

Scientists study many animals to learn more about why and how some animals hibernate. These studies might help humans. Such studies might help people to recover from starvation or hypothermia. Hypothermia occurs when a person's body temperature becomes dangerously low after the person has been in cold weather for a long time. By studying hibernation, scientists may also learn how humans might "hibernate" on long space flights.

◀ *Scientists can put a tag or a special collar on hibernating animals. This helps the scientists to follow the animals for the rest of the year.*

How Can We Help Animals During Winter?

Not all animals hibernate or migrate during winter. You can do many things for animals in the winter. Fill bird feeders or birdhouses with seeds and hang them on tree branches. Spread a mixture of peanut butter and seeds on pinecones and hang them on trees. If you start feeding animals, keep it up through the winter. Animals will start to depend on this food. Never disturb hibernating animals. You put yourself into danger by disturbing them! Visit a nature museum or a forest preserve to learn ways to help animals survive the winter.

Glossary

adaptation (a-dap-TAY-shun) A change that animals make to survive in their environment.

cold-blooded (KOLD-bluh-did) Having a body temperature that changes with the surrounding temperature.

environments (en-VY-urn-ments) The living conditions that make up places.

fuels (FYOOLZ) Powers.

hypothalamus (hy-poh-THA-luh-mus) The part of the brain that controls hunger, thirst, temperature, sleep, and hibernation.

Latin (LA-tn) A language spoken in ancient Rome.

mammals (MA-mulz) Warm-blooded animals that have backbones and hair, and that feed milk to their young.

metabolism (meh-TA-buh-lih-zum) The process which changes food into fuel. It creates the energy the body needs to work.

temperatures (TEM-pruh-cherz) How hot or cold things are.

traits (TRAYTS) Things that make an animal different from other animals.

warm-blooded (WORM-bluh-did) Having a body temperature that stays the same, no matter what the surrounding temperature.

Index

Web Sites

Due to the changing nature of Internet links, PowerKids Press has developed an online list of Web sites related to the subject of this book. This site is updated regularly. Please use this link to access the list:
www.powerkidslinks.com/low/animhibe/